MENTAL MATH SECRETS,
MATH MADE SIMPLE FOR FAST ANSWERS,
IMPROVE YOUR MEMORY

ROGER K. DANETH

Copyright Roger K. Daneth 2013

Disclaimer: By buying, renting, and/or reading this book, the reader agrees that he or she is solely responsible for the results of using any method or technique described in this book to make calculations, estimates, or otherwise implement any information in this book.

Introduction

This book is written for people who need to be fast with basic calculations such as addition, subtraction, multiplication, and division in their daily work, business, or in school.
Often in a dynamic environment, or when you are "on-the-run", there is no time, or it is impractical, to get out the old calculator and punch in digits, when you could do the necessary

math in your head to get the right answer, or at least a good approximation.

This book will teach you with examples of actual calculations for addition, subtraction, multiplication, division, and two methods of obtaining square roots.

The author believes that the techniques shown in this book will also improve your memory with a little bit of practice and due diligence.

So have fun, start learning, and impress your friends and associates with what you can do in your head!

Chapter I

Addition

Of course it is easy to add single digit numbers in your head. Most people know their addition tables, at least for single digit numbers. In the following we will assume that you know the simple grade-school math tables and techniques, so we can start with some real-life examples.

Example 1

Add 54 and 89 in your head. If you can't add the two numbers directly in your head try the following:

First, mentally separate the two numbers into simple combinations.

The sum of 54 and 89 is the same as the sum of the following numbers:

50 + 80 + 4 + 9

Notice, that we have broken the original two digit numbers into two even decade numbers and the two single digit numbers.
Next mentally add the decade numbers and then the single digit numbers to get the result:

50 + 80 = 130

4 + 9 = 13

Then mentally combine the large result with the small result:

130 + 13 = 143

So,

54 + 89 = 143

Summary of mental steps to find result:

54 + 89 = 50 + 80 + 4 + 9

50 + 80 = 130

130 + 13 = 143

This problem only required 3 or less mental steps. Notice that we combined 4 and 9 immediately as an easy mental addition of two single digits.

Now remember, you are not to use pencil and paper to do the above example, or any examples in this book. Try to do the math only in your head. Visualize the numbers as if they were floating just above your eyes in the space in front of you, or in what ever space feels comfortable to your mind.

You might imagine a special place in your mind or outside your mind that is like a scratch-pad where you mentally "see" the numbers "floating" and they stay there until you are done with the calculation or the part of the calculation that you need to remember before you go to the next step. Of course you can easily wipe the "scratch-pad" clean and start over anytime you want to.

Imagine that there is no limit to the size of your mental scratch-pad and you can store as many numbers there as you want to for as long as you want to. Practice doing just that and you will be amazed at what your mind can really do.

Example 2

Now let's try a problem that is a little harder.

Add 569 + 989.

Again, if you can't add the two numbers directly and quickly in your head, try the following:

Breakdown the problem into decade values and single digits:

500 + 900 + 60 + 80 + 9 +9

Combine and add:

500 + 900 = 1400

1400 + 60 = 1460

1460 +80 = 1540   (8 + 6 = 14, so increase 46 to 54 and change 1460 to 1540)

Now add the single digits:

1540 + 9 +9 = 1540 + 18 = 1558.

A faster addition might be to combine 60 and 80 right away to get 140 added to 1400 to make 1540 and then add 18 to get the result 1558. We will do that in our mental math summary.

Notice that in the last step you could have easily originally added 9 + 9 to get 18 and your first breakdown step could have read:

500 + 900 + 60 + 18

This is a good thing to do right away to shorten the string of numbers that you need to remember to do the calculation.

Summary

Let's show the mental steps with the shortened breakdown:

569 + 989

500 + 900 = 1400

1400 + 60 + 80 = 1540

1540 + 18 = 1558

Note that we did not show the breakdown string in the mental math summary. The reason for that is because you already have the breakdown string visible in the problem statement itself, as 5 6 9 + 9 8 9, or visualize the problem as 500 60 9 + 900 80 9.

Also note that in each step of the calculation you only need to remember the original breakdown string of numbers or the problem and the result of the last addition you did. You do not need to remember the intermediate steps, just the last result for the next calculation. This is the technique that you should practice using mentally to speed up your calculations and still be accurate in your final result.

Of course if you can combine any of the steps above and still get the right answer, that is great and you will get answers even faster. Try to jump ahead whenever you can, especially if the problems are simple.

Example 3

Now let's try a 5 digit problem. Again, we will use the same techniques we used above.

Add 14698 to 15999.

With all those 9's it looks like it might be hard but it is not. Let's take it step by step again.

Breakdown the numbers into decade values and single digits:

14000 + 15000 + 600 + 900 + 90 + 90 + 8 + 9

To shorten the list a little try combining the small numbers in the breakdown to make a more compact breakdown:

14000 + 15000 + 600 + 900 + 180 +17

Or even a shorter breakdown with just a little more effort:

14000 + 15000 + 600 + 900 + 197

Start combining:

14000 + 15000 = 29000

29000 + 600 = 29600

29600 + 900 = 30500   (because 6 + 9 = 15, so change 296 to 305 by carrying the 1.)

30500 +197 = 30697

Summary

Let's show just the mental operations with the shortened breakdown:

14698 + 15999

14000 + 15000 = 29000

29000 + 600 = 29600

29600 + 900 = 30,500

30500 + 197 = 30697

In the above summary, why didn't we show the complete breakdown of both numbers?
It's because you can get the smaller numbers as you need them by inspection of the problem

numbers. How did we get the last number to be added, 197? We mentally added the last two digits of each number, 98 and 99 to get 197.

Example 4

Add 1,998,649 to 2,690,349.

This looks like a hard problem but it just has more digits. We will use the same technique as we used above.

We could breakdown the numbers completely to

1,000,000 + 2,000,000 + 900,000 + 600,000 + 90,000 + 90,000 + 8000 + 600 +300 + 49 +49

This is a long list of numbers to remember at one time, so when it comes to large numbers, we will modify the system we use to do more than one breakdown, doing the large numbers first and add, then do the small numbers:

1st breakdown:

1,000,000 + 2,000,000 + 900,000 + 600,000

Combine:

3,000,000 + 1,500,000 = 4,500,000

2ⁿᵈ breakdown and combine:

4,500,000 + 90,000 + 90,000 = 4,680,000

3ʳᵈ breakdown and combine:

4,680,000 + 8000 + 600 + 300 + 49 + 49 = 4,688,998

Summary

Show only the mental steps to solve the problem:

Add 1,998,649 + 2,690,349.

1,000,000 + 2,000,000 + 900,000 + 600,000

3,000,000 + 1,500,000 = 4,500,000

4,500,000 + 90,000 +90,000 = 4,680,000

4,680,000 + 8000 + 600 + 300 + 98 = 4,688,998.

As you can see, only 4 or less steps are required to add the two large numbers with the 'breakdown and combine' method. Also it was easy to combine 49 + 49 immediately to 98 to shorten the string of numbers to add.

Chapter 2

Subtraction

Subtraction is basically addition except that we are adding positive and negative numbers together. Sometimes the result is positive and sometimes the result is negative. Subtraction is no more difficult than positive addition except that you have to keep track of the sign of a number, whether it is positive or negative.

Example 1

Subtract 58 from 69. For ease of mental math we will note the problem as

69 – 58

First breakdown the two numbers into decade numbers and single digits as we did in addition in Chapter 1:

69 – 58 = 60 + 9 – 50 – 8

Re-arrange the breakdown mentally. It will save you time if you do it this way in the first place:

= 60 – 50 + 9 – 8

Combine:

= 10 +1 = 11

Summary of mental steps:

69 – 58 = 60 – 50 + 9 - 8

= 10 + 1 = 11

That was an easy problem. Let's try another.

Example 2

58 – 69 = 50 – 60 + 8 - 9

= - 10 – 1 = - 11

Notice that the problem is the same as Example 1 except that the signs are reserved so the answer is a negative result instead of the positive result in Example 1.

Summary of mental steps for Example 2

58 – 69 = 50 – 60 +8 – 9
-10 – 1 = - 11

Example 3

Subtract 789 from 897:

897 − 789

Breakdown to a string:

800 − 700 + 90 − 80 + 7 − 9

Combine:

100 + 10 − 2 = 108

Summary of mental calculations:

897 − 789

= 800 − 700 + 90 − 80 + 7 − 9

= 100 + 10 - 2

= 108

Example 4

6,567,899, 453 − 6,567,898,354

Breakdown:

6,000,000,000 − 6,000,000,000 + 500,000,000 - 500,000,000 + 60,000,000 − 60,000,000 +

7,000,000 − 7,000,000 + 800,000 − 800,000 + 90,000 − 90,000 + 9,000 − 8,000 + 453 − 354

Notice that most of the above work is wasted effort because most of the numbers cancel each other out except for the following preliminary breakdown after eliminating the numbers that cancel:

9000 − 8000 + 453 − 354

The above is the breakdown we should have noted in the first place by a careful, but quick examination of the problem. In the problem all the numbers are the same until you get to the last 4 digits of the problem, namely 9,453 and 8,354. So the result will actually be the result of the subtraction 9,453 − 8,354.

So now the detail breakdown is:

9000 − 8000 + 400 − 300 + 50 − 50 + 3 − 4

Combine:

1000 + 100 + 3 − 4 = 1099

Summary of the mental math for Example 4:

6,567,899, 453 − 6,567,898,354

9000 − 8000 + 400 − 300 + 3 − 4

1100 – 1 = 1099

In Example 4 the problem looked difficult but it was really a simple matter of comparing the two numbers and mentally noting the differences that do not cancel out. This problem is a typical situation whenever one large number is subtracted from another large number that is approximately equal to it.
After eliminating the numbers that cancel each other, break down the remaining differences and combine to get the result. So, careful examination of the problem allowed mental solution in only two steps.

Chapter 3

Multiplication

You learned a method in grade school on how to multiply. Now we will use a different method that is more suitable for mental math. A simple example follows:

Multiply 3 X 7 :

3 X 7 = [1 + 2] X [3 + 4]

= 1 X 3 + 1 X 4 + 2 X 3 + 2 X 4

= 3 + 4 +6 + 8 = 21

You can do the multiplications in the second line above in any order you like as long as each number in the first brackets is multiplied by each number in the second brackets or vice versa. We will use the simple technique above in all of our two-number multiplication problems that have multiple digits. We will not attempt to prove why the above system works, but you can try any number combinations you want to and the method will always work.

Example 1

Multiply 47 X 69.

First expand the number into the bracket form. We will no longer show the brackets as it is understood how the multiplication of the numbers is performed. We will use : X : to indicate that each of the numbers preceding are to be multiplied times each of the numbers in the following group. We will use X : to indicate that the preceding number is to be multiplied times all of the following numbers. Just X by itself indicates the ordinary multiplication of two numbers together. After the multiplication products are found they are then summed to get the final result.

40 + 7 : X :   60 + 9

= 40 X 60 + 40 X 9 + 7 X 60 + 7 X 9

= 2400 + 360 + 420 + 63

= 2760 + 420 + 63

= 3180 + 63

= 3243

We could shorten the work a little if we combined 420 + 63 = 483 right from the start. Then we would have;

=2400 + 360 + 483

= 2760 + 483

= 3243

The 2nd step just gets a little more difficult. But if you practice you will learn to take the "short-cuts."

Summary of Example 1:

47 X 69

= 40 + 7 : X : 60 + 9

= 2400 + 360 + 420 + 63

= 2760 + 420 + 63

= 3180 + 63

= 3243

In the above summary, we left in all of the little steps, but try to combine steps whenever you can to speed up your calculations.

Example 2

Multiplying larger numbers is of course more difficult, assuming the numbers are not easy decade values, but let's try multiplying two 3-digit numbers:

Multiply 968 X 579 :

Breakdown to simplify:

900 + 68 : X :   500 + 79

1st multiplication and combine (showing here in detail.)

900 X 500 = 450,000

450,000 + 900 X 79

= 450,000 + 900 X : 70 + 9

= 450,000 + 63,000 + 8100

= 513,000 + 8100

= 521,100

2nd multiplication and combine to first sum (in detail.)

521,000 + 60 X 500 + 8 X 500

= 521,100 + 30,000 + 4000

= 555,100

3rd multiplication and combine to second sum (in detail.)

555,100 +   60 + 8 : X : 70 + 9

=555,100 + 4200 + 540 + 560 + 72

= 559,300 + 1100 +72

= 560,472

Summary of Example 2

968 X 579

$900 + 68 \quad : X : \quad 500 + 79$

$= 450,000 + 900 \ X : \quad 70 + 9$

$= 450,000 + 63,000 + 8100$

$= 521,100$

$521,000 + 60 \ X \ 500 + 8 \ X \ 500$

$= 521,100 + 30,000 + 4000$

$= 555,100$

$555,100 + \quad 60 + 8 \quad : X : \quad 70 + 9$

$= 555,100 + 4200 + 540 + 560 + 72$

$= 560,472$

In the above summary I have not shown all of the details, assuming that you will be able to mentally add the easy decade numbers and the small numbers together quickly. Practicing quick additions is one of the secrets to doing fast mental math. Also when you study the above summary, look for steps that you can combine to shorten the amount of work you have to do. The more concise your mental steps can be, the faster you will find the result.

Example 3

In this example we will show another method to use for multiplication. I use the combination X : to mean multiply the preceding number times all the following numbers on the line.

Multiply 5978 X 6578.

Breakdown the second number first.

5978 X : 6000 + 500 + 70 + 8

Now separate the first number in the problem into decade values and multiply one decade number at a time against the breakdown of the second number in the problem:

5978 = 5000 + 900 + 70 + 8

1st multiplication:

5000 X : 6000 + 500 + 70 + 8

= 30,000,000 + 2,500,000 + 350,000 + 40,000

= 32,890,000

2nd multiplication:

Combine previous result with new sum;

32,890,000 + 900 X :   6000 + 500 + 70 + 8

= 32,890,000 + 5,400,000 + 450,000 + 63,000 + 7,200

Combine numbers:

=32,890,000 + 5,850,000 + 70,200

= 32,890,000 + 5,920,200

= 38,810,200

3rd multiplication:

38,810,200 + 70 X :   6000 + 500 +70 + 8

= 38,810,200 + 420,000 + 35,000 + 4,900 + 560

Combine numbers:

= 38,810,200 + 455,000 + 5,460

= 38,810,200 + 460,460

= 39,270,660

4th multiplication:

39,270,660 + 8 X :   6000 + 500 + 70 + 8

39,270,660 + 48,000 + 4000 + 560 + 64

Combine numbers:

39,270,660 + 52,624

= 39,323,284

Admittedly, the above example is a difficult one to do mentally, but with practice you may find that you can do problems like Example 3. In most cases the method of Example 3 is the best method of the two methods I am showing you in this book for multiplying large numbers.
The main trick in doing large number problems is to remember the number breakdowns and strings, and keeping track of intermediate sums as you work. This is the type of mental calculation that you should practice daily to build proficiency and improve your memory.

Summary of Example 3

5978 X 6578.

5000 X :   6000 + 500 + 70 + 8

= 30,000,000 + 2,500,000 + 350,000 + 40,000

= 32,890,000

32,890,000 + 900 X :   6000 + 500 + 70 + 8

= 32,890,000 + 5,400,000 + 450,000 + 63,000 + 7,200

= 32,890,000 + 5,920,200

= 38,810,200

38,810,200 + 70 X :   6000 + 500 +70 + 8

= 38,810,200 + 420,000 + 35,000 + 4,900 + 560

= 38,810,200 + 460,460

= 39,270,660

39,270,660 + 8 X :   6000 + 500 + 70 + 8

=39,270,660 + 52,624

= 39,323,284

The method used in Example 3 is a straightforward and powerful method of multiplying two large numbers together when they are not easy decade values.
But what if you had to multiply some numbers that are special in the sense that they are decade values or partially decade numbers?

Let's do a few simple examples and you will see the ease that such numbers can be computed mentally.

1) 56 X 90:
= 90 X 56
= 90 X : 50 +6
= 4500 + 540
= 5040

2) 800 X 980
= 800 X : 900 + 80
= 720,000 + 64,000
= 784,000

3) 405 X 640
 400 X : 600 + 40
= 240,000 + 16,000
= 256,000
256,000 + 5 X :   600 +40
= 256,000 +3,000 + 200
= 259,200

In the above three examples you will note that numbers with zeros in them are always easier to work with. I call these kinds of numbers with zeroes in them, low density numbers. I call numbers without zeros in them high density numbers.
Another factor that affects the difficulty of mental calculations is whether the numbers have high digits or low digits.

For example a number like 6789 is a high digit number consisting of digits greater than 5. I call numbers like 12345 a low digit number, because all the digits are 5 or smaller, and mental calculations are easier with low digit numbers. But don't let high digit numbers scare you. In this chapter we worked mainly with high density and high digit numbers.

There are also occasions when we do not need an exact answer, but we need a good approximation of the answer.
Here are some examples of problems where approximations are done:

A.) 798 X 897 is approx. equal to 720,000 (8 X 9 =72)
The actual precision answer is 715,806. The error in approximation is less than 1%.

B.) 59,986 X 48,988 is approx. equal to 2,940,000,000 (6 x49 =294)
The actual precision answer is 2,938,594,168. The error in approximation is less than 0.5%. How many zeros should be in the answer? The number of zeroes should be equal to the total number of digits remaining in the problem after the digits you have used to get the first two or three digits of the answer. In this case you used 3 digits, 6 and 49 to obtain the first 3 digits, 294, so there are a total of 7 digits remaining which will

be 7 zeroes following 294 in your answer, or 2,490,000,000.

Depending on the actual numbers and their density, you will generally have to multiply more of the most significant digits (the digits starting from the left) of each number to get higher accuracy estimates.

Chapter 4

Division

A simple example of mental division is as follows.

Divide 57 by 9:

The number 9 will divide into 57, 6 times with a remainder of 3, because 9 X 6 is 54 which is less than 57 and because the remainder of 3 is less than the divisor 9. So 6 will be the first and only digit to the left of the decimal point.

So far we have

57 / 9 = 6 + 3r

The symbol / is the traditional symbol in algebra to indicate than the preceding number is divided by the following number. (We will use it because it is easy to type in our calculations.) We attach the small case letter r to the numerical remainder to identify it as the remainder, in this case 3r indicating that the remainder is 3.

Now to continue the division we add a zero to the remainder to get 30, and then divide this number by 9 to get the next number of the answer which will be the first number to the right of the decimal point.

30 / 9 = 3 + 3r

So far the result of the division is

6.3

Next we again add a zero the new remainder of 3 and divide again.

30 / 9 = 3 + 3r

So now we know that the result is a number with a repeating decimal number, or the result is

57 / 9 = 6.333333 …

To summarize the problem for mental solution it will look like the following:

57 / 9 = 6 + 3r
30 / 9 = 3 + 3r

57 / 9 = 6.3_
30 / 9 = 3 + 3r
57 / 9 = 6.333333 …

Now let's do a more difficult example of division.

Example 1

Divide 89 by 99, or the problem is written as

89 / 99 =

We know that 99 will not go into 89 even once so there will be a zero as the first and only number to the left of the decimal point. So far we have the partial result

89 / 99 = 0. + 89r

Next, add a zero to 89 to make 890 and divide by the divisor 99. 99 will go into 890 almost but not quite 9 times. Why? We can mentally estimate how many times 99 will go into 890 by multiplying 99 by 9 to get 891, and getting a result just one greater than 890. So 99 will "go into" 890 only 8 times. So 8 will be the first number to the right of the decimal point.

89 / 99 = 0.8 + _r

Next mentally multiply the result 8 back times the divisor 99 to obtain 792. Mentally subtract 792 from 890 to obtain the remainder:

8 X 99 = 792
890 – 792 = 98

So the result is now

89 / 99 = 0.8 + 98r

Next add a zero to the remainder 98 to get 980 and mentally divide 980 by the divisor 99. 99 will go into 980, 9 times, so 9 is in the next decimal place following 8.

89 / 99 = 0.89 + _r

Mentally multiply 9 times the divisor 99 and subtract from 980 to get the remainder.

980 – 891 = 89

So now we have

89 / 99 = 0.89 + 89r

Now we already know that with a remainder of 89 we will attach another 0 to get 890 and 99 will divide into 890, 8 times with a remainder of 98. So we have

89 / 99 = 0.898 + 98r

Since we had the same remainder of 98 before, we already know the next decimal place will be filled with a 9 and the remainder will be 89 again. So we have the result with a repeating decimal.

89 / 99 = 0.89898989 …

Let's summarize our mental calculations:

Summary of Example 1

89 / 99 = 0.8 + 89r

8 X 99 = 792

89 X 10 = 890

890 – 792 = 98

89 / 99 = 0.8 + 98r

89 / 99 = 0.89 + _r

9 X 99 = 891

98 X 10 = 980

980 – 891 = 89

89 / 99 = 0.89 + 89r

89 / 99 = 0.898 + 98r

89 / 99 = 0.89898989 …

The above example demonstrates the technique of using mental math to find the remainder of each division, and then adding a zero to it to form the number for the next division by the divisor. Each successive division and calculation of the remainder builds the answer until the complete result is found, or the required degree of accuracy is achieved.

Example 2

Divide 199,200 by 600.

199200 / 600

600 will not divide into 199 so go to the next decade.

1992 / 600 = 3 + _r

600 X 3 = 1800

1992 − 1800 = 192

1992 / 600 = 3 + 192r

192 X 10 = 1920

1920 / 600 = 3 + _r

1920 -1800 = 120

1920 /600 = 3 + 120r

120 X 10 = 1200

1200 / 600 = 2 + 0r

So we now have the result:

199200 / 600 = 332

Summary of Example 2

199200 / 600

1992 / 600 = 3 + _r

600 X 3 =1800

1992 – 1800 = 192

1992 / 600 = 3 + 192r

192 X 10 = 1920

$1920 / 600 = 3 + \_r$

$1920 - 1800 = 120$

$1920 / 600 = 3 + 120r$

$120 \times 10 = 1200$

$1200 / 600 = 2 + 0r$

$199200 / 600 = 332$

Example 2 is a division problem with an exact result because in the last division step there is a zero remainder. An exact result can be a whole number or a decimal number, or a whole number with a decimal, depending on the problem.

Negative numbers can also be divided by a positive number or another negative number. A negative number divided by a positive number or vice versa always has a negative result when dealing only with real numbers.

A real negative number divided or multiplied by a real negative number has a positive result. We will not discuss imaginary numbers as the subject is beyond the scope of this book.
Suffice it to define an imaginary number as a number that is multiplied by or contains a

number that is multiplied by the square root of negative one.

This leads into our next chapter where we will show two methods of obtaining square roots of real numbers.

Chapter 5

Square Roots of Real Numbers

There are a number of ways to calculate square roots. We will discuss two method of calculating square roots that are suitable for finding the square root of numbers that are larger than two digits with mental math.

The recursive square root method

The recursive method is the easiest to use (when doing mental math) for large numbers when only an approximate result is required.

In the recursive method we find the square root of some number Z by making a guess G of the square root and then mentally doing a simple calculation to find a better value for the next guess G. When we have calculated the next value for G, we will use the closest decade value to simplify the next calculation until we have done enough iteration to be close to the actual square root. I will show the basic formula here but you can learn how to do the calculations from our numerical examples that follow.

Z = object number to find the square root of.
G = guess at the square root of Z.
Gc = the calculated value of the next guess to try.
Ga = the actual guess you choose to use in the next calculation.

$$Gc = [\,G + Z/G\,]/2$$

So let's do a numerical example to see how it works.

Example 1

Let the object number be

$Z = 100{,}000$

We want to find the square root of 100,000.

Make a guess at the square root.

$G = 100$, our first guess.

Calculate:

$Gc = [\,100 + 100{,}000/100\,]/2$

$G_c = [\ 100 + 1000\ ] / 2$

$G_c = 1100 / 2$

$G_c = 550.$

So the first calculated next guess at the square root of 100,000 is 550. Instead of 550, let's use an easier number to do the next calculation. Let's use 500 instead of 550 to calculate the next guess.

$G_a = 500$

$G_c = [\ 500 + 100,000 / 500\ ] / 2$

$G_c = [\ 500 + 1000 / 5\ ] / 2$

$G_c = [\ 500 + 200\ ] / 2$

$G_c = 350$

Let's use 300 instead of 350 as the next guess.

$G_c = [\ 300 + 1000 / 3\ ] / 2$

$G_c = [\ 300 + 333\ ] / 2$ ignoring the decimal following the 333.

$G_c = 633 / 2 = 316$ again ignoring the decimal following 316.

We could do more iteration but we are already close to the actual square root of 100,000 which is 316.227766…

Our result in only three iterations of the recursion formula is within 0.1 % of the actual value, even though we did not use the exact calculated values for each new guess, except for the last one, 316. How do you know when to quit? You know when the next guess you calculate is close to the last value you guessed. Then it is time to stop, or you will have to use more difficult non-decade values in your next calculations, so mental math gets a lot more difficult.

By the way if you knew that the square root of 10 is 3.16227766…

Then you could also quickly state the value of the square roots of the following decade multiples of 10:

1000	31.6227766…
100,000	316.227766…
10,000,000	3162.27766…
1,000,000,000	31627.766…

And so forth for every multiple of 10 that has an even number of zeros following 10.

Also, remember that you can test any possible square root of a number simply by multiplying the number by itself, e.g., the square root of 10,000 is 100 because 100 X 100 = 10,000.

Summary of Example 1

Z = 100,000

G = 100, our first guess.

Gc = [100 + 100,000 / 100] /2

Gc = 1100 / 2

Gc = 550.

Ga = 500, new guess

Gc = [500 + 100,000 /500] /2

Gc = [500 + 200] /2

Gc = 350

Ga = 300, new guess

Gc = [300 + 1000 /3] / 2

Gc = 633 /2 = 316 (approximately.)

Sq. root of 100,000 = 316 (approximately.)

Example 2

Let's try a little harder example with the recursive technique again. This time we will just show the numbers in our calculations. I will add the * to the calculation to indicate that I have rounded off the number or dropped decimal values to get an approximation.

Find the square root of 5,267,895.

Guess 1000:

[1000 + 5267895 / 1000] /2

[1000 + 5268] /2 *

= 6268 / 2 *

= 3134 *

New guess 3000:

[3000 + 5267895 / 3000] /2

[3000 + 5268 / 3] /2 *

[3000 + 1756] / 2 *

4756 / 2 = 2378 *

New guess 2000:

[2000 + 5268 / 2] / 2 *

[2000 + 2634] / 2 *

4634 / 2 = 2317 *

Now because the last calculated "new guess" is not much different from the previous calculated "new guess" we are pretty close to the actual square root and we will stop and claim that the square root of 5267895 is approximately 2317.

The actual square root of 5267895 is 2295.1895535..., so we are within 1% of the real square root.
Since we are not really sure of the 3rd and 4th digits in our answer, it would have been smarter to claim that the approximate square root is 2300. With this answer we are within 0.21 % of the correct answer for the square root.
Again, only 3 iterations of the calculation were required to get very close to the actual value.

Summary of Example 2

Find the square root of 5,267,895.

Guess 1000:

[1000 + 5267895 / 1000] /2

[1000 + 5268] /2 *

= 6268 / 2 *

= 3134 *

New guess 3000:

[3000 + 5267895 / 3000] /2

[3000 + 5268 / 3] /2 *

[3000 + 1756] / 2 *

4756 / 2 = 2378 *

New guess 2000:

[2000 + 5268 / 2] / 2 *

[2000 + 2634] / 2 *

4634 / 2 = 2317 *

Square Root the Traditional Way

If we need to find square roots of smaller numbers where we need to get the decimal digits of the square root as well as the whole number part, then it is better to use one of the traditional methods of calculation. The object here will be to simplify the process to the bare bones so that mental calculations are possible. Admittedly, this will be more difficult than the previous method, but with practice you should be able to master it.

Example 1

Find the square root of 899.

Group the numbers in pairs of two from the right:

8 99

Find a number whose square is less than the first number in this case 2. Mentally float this number above the first number (in the result line):

2
8 99

Double the floating number and submerse it below the first number:

2
8 99
4

Subtract:

2
8 99
-4
-----

4 99

Double top floating number and float it below and to the left of the result with a space to the right of it:

```
        2
        8 99
        -4
        -----
4_      4 99
```

Add a number to the right of 4 in the space provided such that when the new number is multiplied by the added digit, the result will be less than the remainder of the previous subtraction and do the new subtraction so formed. Put the added number above, floated in the result line. Add the decimal point after the new number in the result line:

```
          2 9.
          8 99
          -4
          -----
49        499
49X9=    441
          -----
           58
```

Now we are done with the first subtraction so you can forget about it and simplify the problem. Next bring down the next pair of digits (two zeroes) after the decimal point following 899, making 5800 as the new remainder to work with. Float 58 to the left of the remainder:

```
           2 9 .
           8 99. 00
58_          58  00
```

Now find a number to add to 58 that will multiply itself to a number less than 5800. It will be 9. Place the 9 in the result line and multiply 589 X 9 = 5301. Subtract 5301 from the last remainder of 5800 to find the next remainder:

```
              2 9 . 9
              8 99.00 00
589            58 00
589X9          53 01
               ------
598_             499 00
```

Now we are done with the 2nd subtraction and we can simplify our problem to show the next step. Bring down the next pair of zeroes to form the next remainder:

```
              2 9 .9
              8 99.00 00
598_             499 00
```

Find the next digit on the result line by finding a digit to add onto 598_ so that it will divide into the next remainder of 49,000. The next digit is 8 because 598 X 8 is less than 49,900:

```
               29 .9 8
           8 99.00 00
   5988        499 00
   5988X8      479 04
               --------
                1996
```

At this point we have found the square root to two decimal places. We could continue the calculation process to as many decimal places as desired, but by this time you know how the process works.

The actual square root of 899 is 29.9833287… The error of our calculation is a little over 0.01% which is a small enough error for most purposes.

Summary of Example 1

Find the square root of 899.

8 99

2
8 99

```
      2
   8)99
      4


      2
   8)99
     -4
    ----
     499


               2
            8)99
              -4
             ----
      4_    499


              2 9.
           8)99
             -4
            ----
     49     499
     49X9 = 441
            ----
              58


              2 9 .
           8)99.00
    58_       58 00


              2 9 . 9
           8)99.00 00
    589       58 00
```

```
589X9    53 01
         -----
598_      499 00

          2 9 .9
         8 99.00 00
598_      499 00

          29 .9 8
         8 99.00 00
5988      499 00
5988X8    479 04
         --------
          1996
```

Chapter 6

Practical Problems

1. The Orient Express™ is to make a run from Moscow to Istanbul, a distance of approximately 1333 miles.
The train gets about 400 miles on a gallon of fuel for each ton of weight.
The fuel costs the railroad about $2 per gallon.
The locomotive weighs about 840 tons and each car weighs about 65 tons empty.
There are 15 cars and each car holds about 40 people.

The railroad company always uses a figure of 250 pounds for each passenger's weight including baggage.

The trip takes 30 hours on the average including stops.

a) How much fuel is required for a one-way trip?
b) What is the total cost of the fuel for the one-way trip?
b) What is the average speed of the train for the trip?
c) If the train is full what should each passenger's ticket cost to just cover the cost of the fuel?

First we need to calculate the total weight of the train in tons. We will assume that the train is full with every seat taken. The mental calculations are shown in detail to be as clear as possible. You can eliminate the obvious steps if you can do the calculations in fewer steps:

W = 840 + 15 X 65 + [40 X 15 X 250 / 2000]

= 840 + 10 X 65 + 5 X 65 + [600 X 250 /2000]

= 840 + 650 + 325 + [150000 / 2000]

= 800 + 600 + 300 +40 +50 + [150 /2]

= 1700 +40 +50 +25 + 75

= 1700 + 190 = 1890 tons.

Now we calculate the amount of fuel required:

Gallons per ton for the trip = 1333 / 400

= [1000 + 300 + 30 + 3] / 400

= 2.5 + 0.75 + 0.075 + 0.0075

Multiply and divide by 10,000 to get rid of the decimals:

= [25000 + 7500 + 750 + 75] / 10000

= [32500 + 750 + 75] / 10000

= [33250 + 75] / 10000

= 33325 / 10000 = 3.3 gallons per ton, approximately.

Total gallons equals gallons per ton, times total train weight in tons;

= 3.3 X 1890

Multiply and divide by 10 to remove the decimal.

= 33 X 1890 /10

= [1890 X : 30 + 3] / 10

= [56700 + 5670] / 10

= [56700 + 5000 + 600 + 70] / 10

= [61700 + 600 + 70] / 10

= [62300 + 70] /10

= [62370] / 10 = 6237 gallons of fuel required.

Total cost of the fuel at $2 per gallon is

= $ 2 X : 6000 + 200 + 30 + 7

= 12000 + 400 +60 + 14

= $ 12,474

(You might have been able to multiply 6237 by 2 directly to get 12474.)

The total number of passengers for a full train with 15 cars and 40 people per car is

= 40 X 15 = 600 people.

The cost per passenger for fuel is

= $ 12474 / 600

Divide by 6:

12 / 6 = 2 + 0r

4 / 6 = 0 + 0r
47 / 6 = 7 + 5r
54 / 6 = 9 + 0r

12474 / 6 = 2079

Divide again by 100:

$ 2079 / 100 = $ 20.79

So each passenger's ticket should cost at least $20.79 just to cover the cost of fuel. Of course the ticket will be much higher when other costs are added such as overhead, maintenance, service, taxes, and profit are also included in the ticket.

The average speed of the train is computed from the total distance of the trip divided by the total time for the trip one-way:

= 1333 miles / 30 hours

= 44. 4 miles / hour

2. A Boeing 747 jet aircraft is to make a trip from New York City to Sydney Australia,
a distance of 9935 miles.
The aircraft burns approximately 1 gallon of fuel per second at 500 miles per hour on the average,

depending on weather conditions, and the load the aircraft is carrying.

a) If the plane is flying at 500 miles per hour and under average conditions burns 1 gallon of fuel every second, how much fuel does the aircraft burn?

b) If the fuel cost is $2 per gallon, how much does a one-way load of fuel cost assuming it is all burned on the trip?

c) If the plane is carrying 300 passengers, how much is the fuel cost per ticket?

d) How long does the flight take?

First calculate how long the flight time is:

= 9935 miles / 500 miles per hour

= 1987 / 100 = 19.87 hours.
Round off to 20 hours.

Now calculate how many seconds there are in 20 hours:

20 X 3600 = 72,000 seconds.

So at 1 gallon per second, 72,000 gallons of fuel are consumed in the one-way trip.

At $2 per gallon the fuel cost is 2 X 72000 = $144,000.

The fuel cost per ticket is then:

$ 144,000 / 300

= $ 480 per ticket.

IMPORTANT NOTE: IF YOU ARE DOING CALCULATIONS WHERE LIVES MAY DEPEND ON ACCURATE RESULTS, MAKE SURE YOU DOUBLE CHECK YOUR RESULTS AND ADD SAFETY MARGINS AT ALL TIMES.

3. A farmer buys a field that has the shape of a right triangle. The farmer plans to graze cattle on the field so he wants to put a barbed wire fence around the property. One side of the right angle sides is 5000 feet and the other side is 2500 feet. He has to calculate the length of the third side of the fence and then order enough barbed wire to put up a fence with 4 strands of wire. The barbed wire costs $0.04 per foot.
a) What is the length of the third side of the triangular field?
b) How much wire does the farmer need?
c) How much is the total cost of the wire?

The third side of the field is the square root of the sum of the squares of the other two sides of the field.

First calculate the squared value of the third side of the field:

$= 2500^2 + 5000^2 = 2500 \times 2500 + 5000 \times 5000$

$= 2000 + 500 : X : 2000 + 500 \quad + 25{,}000{,}000$

$= 4{,}000{,}000 + 2{,}000{,}000 + 250{,}000 + 25{,}000{,}000$

$= 31{,}250{,}000$ feet squared

Now find the square root of 31,250,000 to get the length of the third side of the field in feet. Use the recursive method:

Guess 5000:

[5000 + 31250000 / 5000] / 2

= [5000 + 6250] / 2

= 2500 + 3125 = 5625

New guess 5600:

[5600 + 31250000 / 5600] / 2

= [5600 + 312500 / 56] / 2

calculate 312500 / 56:

$312 / 56 = \mathbf{5} + [312 - 280]r = 5 + 32r$

$325 / 56 = \mathbf{5} + 45r$

$450 / 56 = \mathbf{8} + [450 - 448]r = 8 + 2r$

$20 / 56 = \mathbf{0} + 0r$

Take the whole number part only = 5580.

Now average 5600 and 5580:

[5600 + 5580] /2

= 5590 feet, the length of the third side of the triangle.

A calculator gives the answer as 5590.169944… so our answer of 5590 feet has an error of approximately 0.003 %.

The total length of all the sides of the triangle is then

2500 + 5000 + 5590

= 7500 + 5000 + 500 + 90 = 13,090 ft

4 X 13090 = 52,360

52,360 feet of wire is required for 4 strands.

The cost of the wire is

= $ 0.04 X 52360

= 4 X 52360 / 100

= 209440 / 100

= $ 2094.40

= $ 2094.40 total cost for the barbed wire.

Conclusions

Now that you have read this book, you are ready to start doing mental calculations. It takes lots of practice to develop mental math techniques, unless you are a genius or have a talent for mental math.
Always double check your calculations and if you make errors at first don't feel bad, but go back and find out what you are doing wrong, and make corrections.
You may find better and easier ways to do mental math than what is shown in this book. That is great. But whatever you do, keep practicing. You will find that your memory will improve and you will become faster over time. You will probably

find short cuts that you can do to speed up your calculations.

Finally, if life and limb is at stake, make sure your calculations are correct. It might even be a good idea to check your work with a calculator or a computer.

Good luck and happy calculating.

www.ingramcontent.com/pod-product-compliance
Lightning Source LLC
Chambersburg PA
CBHW030508220526
45464CB00006B/2713